Acknowledgments

Sulwhasoo would like to thank the founder, Suh Sung-hwan, for opening the path to the secrets of Asian beauty. Thanks are also due to Professor Jean-Noël Kapferer for his support and branding guidance. Finally, this book would not have been possible without the helpful contributions of photographers Koo Bohnchang, Bae Bien-u, Min Byung-hun, artists Song Hoon and Bae Se-hwa, potter Kang Suk-young, the Kansong Art Museum, and the AMOREPACIFIC Museum of Art.

A treatment room in the Sulwhasoo spa in Seoul, South Korea, which opened in October 2008. Services conclude with breathing routines and tea. Courtesy of Sulwhasoo.
Sammi brand products. Amore Sammi was the version sold in Korea; True Sammi was the version sold overseas. Courtesy of Sulwhasoo.

Hydro-aid Moisturizing Lifting Serum, Hydro-aid Moisturizing Lifting Mist, and Hydro-aid Moisturizing Lifting Cream. The Hydro-aid line contains botanical extracts that soothe sensitive skin, revitalize moisture flow to firm and brighten, and relieve stress to comfort the skin. Courtesy of Sulwhasoo.

Herblinic Restorative Ampoules. These highly concentrated Korean herbal medicinal ampoules quickly restore aggravated skin. Courtesy of Sulwhasoo.

Holly Siegel

雪花秀

Sulwhasoo

ASSOULINE

電記福

The challenge of achieving balance in all things is universal—we try to harmonize every aspect of our lives, from our budget to our time. The beauty of nature comes from the perfect equilibrium of principles—light and dark, hot and cold, wet and dry. Beautiful skin, too, requires maintaining a fragile balance of elements.

Sulwhasoo skin care, one of the most popular lines in Korea, was created with a deep understanding and respect for consonance in the skin, in nature, and in oneself. Following the traditional Korean tenet of *sang-seng*—the ultimate harmony between opposite energies—Sulwhasoo marries the elegant wisdom of nature to novel scientific advances. The result is an effective and luxurious collection of skin care products that is finally being enjoyed by women outside Korea.

The folklore of Korea is full of tales that underscore the requirement of unyielding selflessness in order to exude the delicate gracefulness prized by the culture. The story of the Fountain of Youth, in which the rejuvenation of a married couple after sipping from a magical fountain incites jealousy in a neighbor who then bathes in the water and becomes a helpless infant, highlights the dangers of excess and vanity. Nothing exists without its opposite, so no extreme is the ideal position. The benefit of restraint is a synergy of mind, body, and spirit.

Korean culture has always celebrated feminine beauty with a marked consideration for humility and depth. In fact, for a woman with a less-than-pure heart to appear beautiful has historically been considered a serious act of deception. Complete beauty involves a delicate balance of attributes. Still, tomb paintings from ancient times hint at the use of cosmetics by Korean women as far back as the Three Kingdoms period (57 BC to 668 AD). A beauty guide for upper-class Korean women from 1809 called the Kyuhapch'ongso details techniques for making cosmetics and describes multiple popular eyebrow shapes, including one shaped after a willow leaf.

Women in modern Korea are among the most concerned with beauty products and treatments in the world, though they differ from many other Asian markets by maintaining a strong commitment to using Korean-made products. The average Korean woman is said to apply about seventeen products on her skin every day. In America, skin care is typically offered in a three-step system—cleanse, tone, moisturize—while Korean skin care typically offers a four-step system of refiner, emulsion, essence, and cream. Korean women are very protective of the whiteness of their skin, employing whitening creams, parasols, and hats to guard against hyperpigmentation.

The story of Sulwhasoo begins in the lush mountains surrounding the small Korean city of Gaeseong, through which the Imjin and Yesong rivers flow. In Gaeseong the most potent ginseng has been harvested since the Goryeo dynasty (918-1392). Gaesong is also the birthplace of Suh Sung-hwan, the founder of AMOREPACIFIC skin care.

Suh's mother, Madame Yoon Dok-jung, began making and selling her own high-quality camellia hair oil in 1932. Hair oil made from camellia or wild sesame seeds was popular with Korean women (the AMOREPACIFIC Museum of Art has a whiteware bottle bearing a nineteenth-century dragon character determined to have housed the oil), as one of the main focal points of their aesthetic in that period was the maintenance of fragrant, long, dark hair. The preoccupation with hair was nothing new: In eighteenth-century Korea, the most popular hairstyle required a wig made of braids that women so coveted that it was outlawed by a decree from King Chongjo in 1788, who believed the frenzy for them was against Confucian values.

A young Suh accompanied Madame Yoon on her trips to the Gaeseong markets in search of the highest quality ingredients and developed a keen interest in and knowledge of traditional Korean herbs. The medicinal and cosmetic advantages of ginseng in particular made a lasting impact on him.

The women of Korea have long taken advantage of the beautifying powers of ginseng. During the Joseon dynasty, brides prepared for their weddings by bathing in it so that their bodies would be detoxified and their complexions brightened and clarified, and ginseng root is also said to have been included in facial scrubs. In Korean herbal medicine, ginseng tea was also used to prevent signs of aging, skin discoloration, and blemishes.

By 1945 Suh had been studying the principles of sang-seng and became interested in combining the philosophy with his interest in indigenous ingredients and their health and beauty benefits. In October of 1972, after years of research and development, the company he founded, AMOREPACIFIC, patented potent ginseng extracts. They launched Ginseng Sammi, the first cosmetics line in Korea to include ginseng saponin, one year later. In the years that followed, AMOREPACIFIC has continued to develop ways of incorporating the ideas of sang-seng with the efficacy of ginseng and other natural ingredients.

The brand's devotion to Eastern medicinal ingredients is evident in the development of its original formula: Researchers at AMOREPACIFIC pored over books from Korea, China, and Japan to develop a master list of herbal ingredients with skin care benefits. After editing the final list of over 500, they created Sulwha, Sulwhasoo's predecessor, in 1987.

In 1997 AMOREPACIFIC introduced what would become Korea's premier skin care brand, Sulwhasoo. The company funds institutions like the AMOREPACIFIC Museum of Art in order to preserve the traditional Korean culture, especially as related to women's cosmetics, to give brands including Sulwhasoo a rich bounty of information to draw upon. This is how Sulwhasoo strives to be not only a product line but also a way to spread awareness of Korea's rich heritage.

The ingredients in Sulwhasoo are very rare and not commonly found in skin care products. For example,

Hedyotis diffusa, a native herb noted for its high purity levels (three times stronger than vitamin C) and used for cooling and detoxifying, is found exclusively in Sulwhasoo products. Sulwhasoo has developed two components to supplement and protect the yin energy in skin, a process referred to as Jaeum. The first was Jaeumdan, an exclusive blend formulated with five Korean herbal medicinal plants: small Solomon's seal, adhesive rehmannia, white lily, Chinese peony, and East Indian lotus. Each ingredient is integral in the maintenance of flawless skin.

The root of small Solomon's seal, dried in sunlight, provides intensive hydration to the skin. This plant, a close relative of lily of the valley, has been celebrated as a miracle plant by healers all over the world for its curative properties.

Adhesive rehmannia root is valuable for balance in the body and skin, increasing fluid circulation and bringing energy to the five internal organs. Beyond its outstanding moisturizing capacity, the root also prevents dullness and pigmentation.

White lily bulb clarifies the body by helping to distribute energy and fluids throughout the body, providing a sense of tranquillity and refining the skin. The perennial flower contains polysaccharides, which help keep water in the skin. In ancient times, white lily was brewed to cleanse wounds and treat burns.

Chinese peony root, also sun-dried, contains tannin and paeoniflorin, both known to bestow serenity and anti-inflammatory benefits. Chinese peony heightens yin in the liver and stimulates blood circulation, preserves the internal organs, and boosts immunity.

The mature seed of the East Indian lotus replenishes energy lost in the skin, which allows for moisturization and even tone. This evergreen plant grows underwater, with breathtaking, sweet-smelling pink flowers blooming above the surface. East Indian lotus is native to the warm parts of Asia and Australia.

In 2004 Sulwhasoo introduced Jaeumboweedan, which took the Jaeum concept even further. The updated formula was fortified with kobon root to soften the skin, mountain peony root to calm skin with the power of paeonol, and Adlay millet to nourish skin with its high concentration of protein and amino acids. Two additional ingredients were included to help balance the formula and provide additional benefits: dwarf lilyturf root and licorice root, the latter said to bear the nine energies of the earth and to be capable of synergizing seventy-two minerals and twelve hundred herbs.

The unique way that the raw ingredients in Sulwhasoo are manipulated, referred to as the Poje Method, treats each component in the manner that maximizes its healing and beautifying benefits. They may be steamed, fermented, baked, or enriched. Where other brands may only use one part of a plant, Sulwhasoo takes its cue from nature and always includes ingredients in their entirety—leaf, stem, and root.

With the Poje Method, Sulwhasoo skin care utilizes honey's ability to harmonize hundreds of medicinal herbs and maximize the effect of ingredients. Small Solomon's seal and white lily—important components of the original Jaeumdan formula—as well as ingredients added in Jaedumboweedan, Satsuma mandarin, ladybells, kobon, and Adlay millet are steeped in honey and baked in earthenware jars to increase their power. This time-honored traditional method is called the Korean Milja Process.

Treating other ingredients with salt, particularly those in Sulwhasoo's Snowise line, which brightens and refreshes skin, is another part of the Poje Method. Logs from the mulberry tree, known to be a strong whitening agent, are immersed in saltwater and baked. Salt has clarifying and fortifying properties in the blood, benefits Sulwhasoo extends to the skin to balance the yin energy and remove dead skin cells, revealing more radiant skin.

A star product of the Sulwhasoo line is the Concentrated Ginseng Cream, which uses various parts of the plant. Formulated with the roots, berries, and water of rare Korean ginseng, the blend is synonymous with antiaging care in Korea. Steaming the roots for two to three hours then drying them in the sun for two or three days yields red ginseng, which contains saponin—an ingredient that regenerates the skin.

Sulwhasoo's Timetreasure line revives dull complexions and gives skin nutrients and energy it may be lacking, through a method of harvesting ingredients called the Germinating Treatment. Five different thoroughly ripened seeds and fruits are germinated at a specific temperature, then two of these ingredients—red beans and black sesame—are steeped in hot spring water for four to eight hours. Thanks to this time-tested holistic approach to extracting the strongest and most potent energy from the ingredients, the amino acids and glutamine in the red beans and the black sesame are even more effective in strengthening the natural barrier of the skin. The Prunus mume flower, one of the Four Gracious Plants of Korea (the others are chrysanthemum, orchid, and bamboo), was a beloved symbol used by Korean writers who prized its depth, as its breathtaking, fragrant flowers bloom deep within the mountains, far from the human gaze. The flower has been used as a metaphor for the subtlety in beauty that Korean women strive for—it is even believed that the scent of the Prunus mume is so pervasive and all-encompassing that it is audible.

The Men's Care line has its own signature ingredient base, known as Jeongyangdan, which includes Japanese cornelian cherry (a type of dogwood) and red mushroom extract. These are steamed in alcohol, a method long practiced in Korean herbal medicine to improve the circulation of the ingredient's energy. The Japanese cornelian cherry and red mushroom are steeped in Andong soju, sealed airtight until thoroughly absorbed, and then steamed. Another crucial component of the Men's Care line is pine, essentially the national tree of Korea, used as a symbol of integrity (in the Joseon dynasty, scholars used its scent both on their person and in the form of incense to convey sophistication) and for furniture.

Sang-seng is achieved through the balance of yin and yang energies, which results in an all-encompassing harmony that radiates through the skin. Korean herbal medicine designates a seven-year cycle for female changes both internal and external. The fifth cycle, which occurs at age thirty-five, brings the first signs of aging: a dull and dry complexion. At forty-two, a woman enters the sixth cycle, when additional changes occur, such as the graying of the hair and an even stronger need for the type of multidimensional skin care Sulwhasoo provides. By singling out these symptoms and developing the most targeted treatments and remedies found in nature and enhanced by science, Sulwhasoo quickly became integral to the beauty regimen of women throughout Korea.

Before long, Sulwhasoo's flagship product, First Care Serum, was the best selling of its kind in Korea, and the company has been happy to give back to the country that both inspired and supported it. In 2003 Sulwhasoo announced the formation of the Sulwha Club, which comprises a number of forward-thinking creative luminaries including designers, artists, directors, photographers, writers, and sculptors. For its annual Sulwha Cultural Night, some of the group's members contribute works that are auctioned to benefit the preservation of Korean heritage and resources. In 2006 eight pieces based on the theme of the Beautiful Colors of Korea were sold to preserve the country's pine trees.

To disseminate the beauty of Korean tradition, the Sulwha Cultural Exhibition was created. The brand strongly supports the arts and crafts in maintaining its heritage, nurturing master artisans who continue to benefit the country for decades. Sulwhasoo acts as a gateway for these artists as well as new, rising talents who reinterpret traditional crafts and artwork in a more modern shape.

Nowhere is the philosophy of sang-seng more apparent than in Sulwhasoo's impressive championing of the arts. If a piece is created simply to be sold for profit, it has no integrity, even if it is a beautiful work; but when it becomes a vehicle for viewers to appreciate and understand the rich heritage of a nation, it becomes more than the sum of its parts. Furthermore, to invest its cost back into the natural resources that provided the artist with inspiration is an example of the type of balance that sang-seng refers to.

In all things, Sulwhasoo skin care is thoughtful and respectful of sang-seng principles. It goes without saying that a company with such an impressive understanding of the importance of aesthetics and tradition would employ highly sophisticated and considered design. Sulwhasoo is packaged in elegant curved bottles reminiscent of ancient Korean ceramics—the shape of Goryeo celadon, the silhouette of the Korean mountains—meant to fit gracefully into the palm of the hand. The original Korean logo was drawn by the famed calligrapher Song Kyung-sik.

The limited-edition ShineClassic compacts have had a cult following since they debuted in 2003. The elegant cases are inspired by traditional Korean craftsmanship, such as the lattice and floral pattern, the look of buncheong and celadon ceramics, and the blossoms of peony and Prunus mume, which have long been the basis of patterns used in Korean palaces and Buddhist temples. The compacts come in a pouch reminiscent of the bok'jumuni, which was historically carried by women during the Joseon dynasty to hold small personal items.

Of course, the legacy of Suh Sung-hwan's mother, Madame Yoon Dok-jung, would not be fully exalted without Sulwhasoo's Hair Treatment Oil. The oil extracted from the seeds of the camellia tree is a traditional component of such hair oil in Korea and across Asia, as it's known to help hair grow, boost its shine, and fortify the scalp. Sulwhasoo Hair Treatment Oil contains over fifty percent natural camellia oil.

In October 2008 Sulwhasoo opened a spa in Seoul, South Korea, a luxurious space with a modern aesthetic. White walls, tiles, and window fixtures house dark wood furniture decorated with traditional lamps and vases. A respect for tradition carries to the treatments as well, wherein cool jade is used to heighten the effects of ginseng, a warm mugwort pad is draped on the lower abdomen, and a dry cloth is infused with Korean aromatherapy oils—a ritual dating back to the Joseon dynasty. Services conclude with breathing routines and tea.

In a world where quick fixes like plastic surgery are on the rise—a phenomenon to which Korea is most certainly not immune—the time-honored techniques and ingredients used in Sulwhasoo's products are as refreshing as the company's commitment to sang-seng. With generations of Korean wisdom paired with modern scientific advances, rooted in a deep consideration for balance in all things, Sulwhasoo is not only a product line but also a lifestyle.

雪花

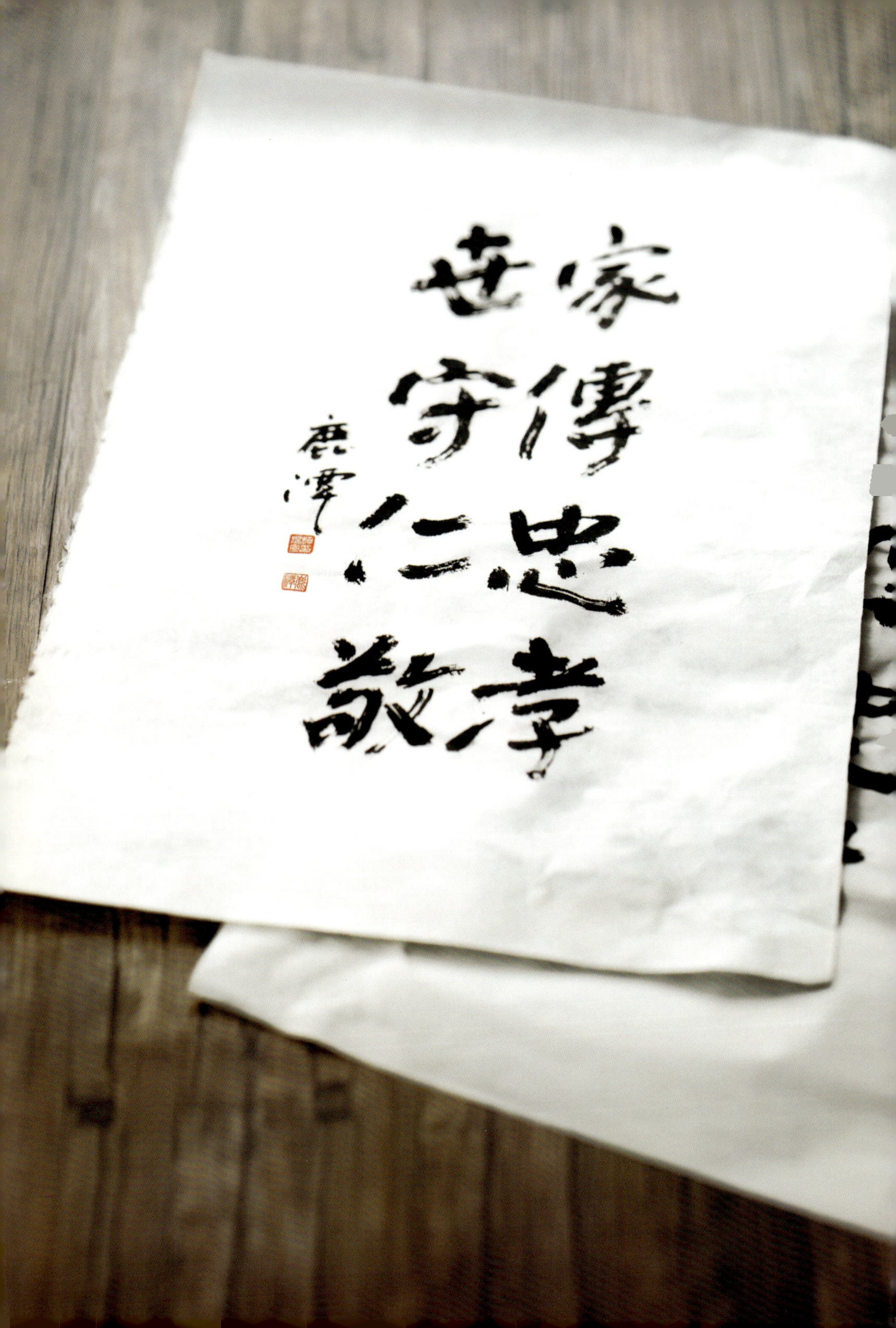
家傳忠孝
世守仁敬
鹿潭

雪
花
秀

Sulwhasoo

HAIR GLAZE OIL
동백윤모오일

" All should be risked for beauty. "

Buddhist proverb

66 Nature is ancient but surprises us all. 99

Björk
Singer-songwriter

> **Renew thyself
> completely each day.**

Henry David Thoreau
Thinker, essayist, and naturalist

66 Choose only one master—Nature. **99**

Rembrandt
Painter and etcher

66 Beauty is eternity gazing at itself in a mirror. But you are eternity and you are the mirror. 99

Kahlil Gibran
Philosophical essayist, novelist, mystical poet, and artist

雪花
秀

윤조(潤燥)에

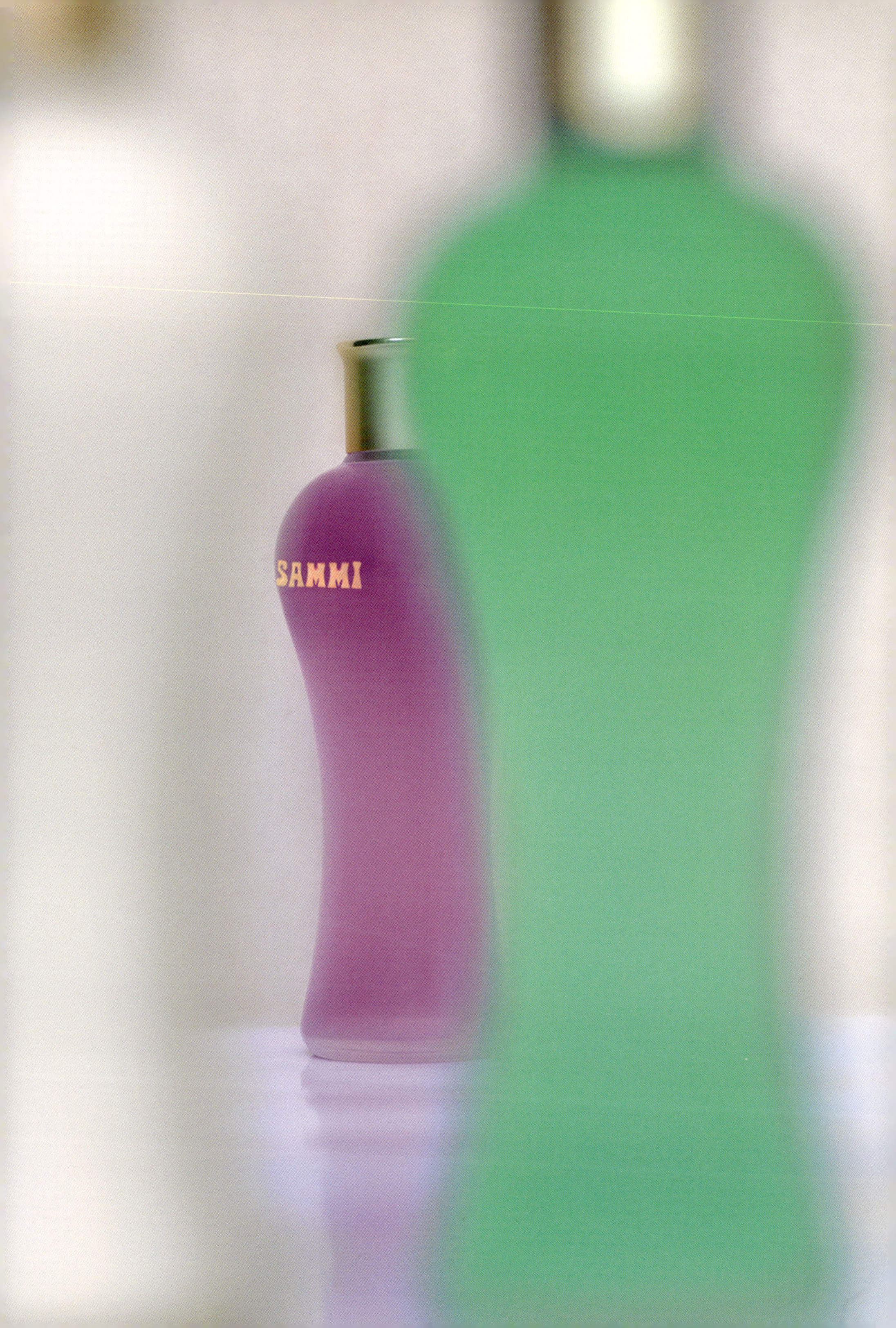
SAMMI

66 Beauty doesn't need ornaments. Softness cannot bear the weight of ornaments. **99**

Munshi Premchand
Hindi-Urdu novelist, story writer, and dramatist

66 Sang-seng is the ultimate
harmony between opposite energies. **99**

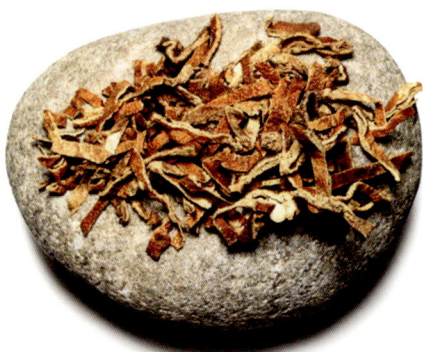

66 Korean beauty of form is one of unifying and harmonizing with space... We inherently aspire to be one with the nothingness of Mother Nature. 99

Yi Oryong
Author and former Korean Minister of Culture

66 Nature, inexhaustible treasure of colors, sounds, of forms and rhythm, unequaled model of total development and of perpetual variation. Nature is the supreme resource. **99**

Olivier Messiaen
Composer

Chronology

1809	The Kyuhapch'ongso, a beauty guide for upper-class Korean women, details techniques for making cosmetics, and describes multiple popular eyebrow shapes, including one shaped after a willow leaf.
1932	Madame Yoon Dok-jung, the mother of founder Suh Sung-hwan, begins making and selling her own high-quality camellia hair oil.
1945	Suh Sung-hwan founds his cosmetics business, combining the principles of sang-seng with indigenous ingredients for their health and beauty benefits.
1967	Begins researching ginseng-based medicinal skin care methods.
1972	After years of research and development, AMOREPACIFIC patents potent ginseng extracts.
1973	Launches Ginseng Sammi, Korea's first cosmetics brand to include ginseng extracts.
1987	Sulwha, Sulwhasoo's predecessor, is created.
1997	AMOREPACIFIC introduces what would become Korea's premier cosmetics brand—Sulwhasoo.
2003	Sulwhasoo announces the formation of the Sulwha Club, comprising creative luminaries including designers, artists, directors, photographers, writers, and sculptors.
	Sulwhasoo makes an impact on Korean culture through its popular eponymous publication, *Sulwhasoo* magazine.
2004	Sulwhasoo introduces Jaeumboweedan, which further develops the Jaeum concept. Sulwhasoo launches in Hong Kong.
2010	Sulwhasoo launches in Bergdorf Goodman in New York City.
2011	Sulwhasoo launches in China.

Sulwhasoo

Sulwhasoo logo drawn by the famed calligrapher Song Kyung-sik.
Courtesy of Sulwhasoo.

Portrait of a young Korean woman in geisha costume, by an anonymous
artist, from 1758. Kansong Art Museum. © Scala/White Images/Art Resource.
Prunus mume (Japanese apricot) lattice pattern, featured in the Sulwhasoo spa
and on counters and outer boxes. This type of lattice and floral pattern has long
been used in Korean palaces and Buddhist temples. Courtesy of Sulwhasoo.

A piece of celadon pottery (16th–17th century) from the AMOREPACIFIC Museum
of Art. Courtesy of Sulwhasoo.
Sulwhasoo products from the 2006 Snowise line, packaged in elegant curved
bottles reminiscent of ancient Korean ceramics, such as Goryeo celadon.
Courtesy of Sulwhasoo.

Spa treatment room in Hong Kong flagship, opened in 2009.
Courtesy of Sulwhasoo.
Honey Treatment. Honey harmonizes hundreds of medicinal herbs and
maximizes their effects. This method not only refines the ingredients but also
effectively refines the skin. Courtesy of Sulwhasoo.

Motto handed down by King Sejong the Great of the Joseon Dynasty.
It states: "In the family, be loyal to the country and serve your parents;
in society, be kind to others and respect the elderly." Courtesy of Sulwhasoo.
Sulwhasoo Hair Treatment Oil restores the natural radiance of hair.
Courtesy of Sulwhasoo.

Special herbal blends and techniques designed to extract the most potent
energy from ingredients. *Clockwise, from top left:* Jaeumdan, jaeumboweedan,
alcohol steam treatment, steam treatment, germinating treatment, salt treatment.
Courtesy of Sulwhasoo.
Ceramics by Lee Yoon-shin, a former member of the Sulwha Club. His ceramics
were displayed in the 2006 Sulwha Cultural Night. Courtesy of Sulwhasoo.

Facade of Sulwhasoo flagship on Canton Road in Hong Kong. Courtesy of Sulwhasoo.

Detail of Sulwha Club member Bae Bien-u's *Pine Trees,* from a series displayed in the 2006 Sulwha Cultural Night. Courtesy of Sulwhasoo. The first Sulwhasoo products, introduced in 1997. The product designs were inspired by traditional aesthetics, realized in the richness of porcelain carefully kneaded by a master craftsman. Courtesy of Sulwhasoo.

The *Prunus mume* flower, one of the Four Gracious Plants of Korean tradition, is a beloved symbol prized for its breathtaking, fragrant flowers that bloom deep in the mountains. © Bae Bien-u.

For centuries, the delicate *Prunus mume* blossom has been a metaphor for the subtlety in beauty that Korean women strive for. Courtesy of Sulwhasoo.

Book art by Sulwha Club member Yoo Rim, whose books and tools were displayed in the 2006 Sulwha Cultural Night. Courtesy of Sulwhasoo.

Close-up of a tree trunk. © Dinodia/The Bridgeman Art Library International.

Painting of women washing laundry, grooming their hair, and entertaining themselves, by Korean artist Sin Yun Bok. Kansong Art Museum. © Scala/White Images/Art Resource.

Sulwha products from 1987. Courtesy of Sulwhasoo. Botanical illustration of the East Indian lotus, a component in Sulwhasoo's exclusive Jaeumdan blend of Korean herbal medicinal plants, intended to replenish energy in the skin. © Song Hoon.

Sulwhasoo Hydro-aid Moisturizing Lifting Serum improves the moisture constitution of the skin and contains seaweed known to discharge impurities, promote circulation, and relieve asthma. Courtesy of Sulwhasoo.
Sulwhasoo White Ginseng Brightening Mask promotes skin's natural circulation and hydration. Ginseng root has long been used for brightening, clarifying, and detoxifying. Courtesy of Sulwhasoo.

Essential line including First Care Serum (Sulwhasoo's "hero" product), Balancing Water, Balancing Emulsion, Rejuvenating Eye Cream, and Revitalizing Serum. This essential regimen promotes balance while replenishing moisture and nutrients. Courtesy of Sulwhasoo.

The elegant cases of Sulwhasoo's limited-edition ShineClassic compacts are inspired by Korean craftsmanship and traditional symbols such as *Prunus mume* blossoms. Courtesy of Sulwhasoo.
The *Prunus mume* flower is a central design motif for Sulwhasoo and its ShineClassic makeup series. Courtesy of Sulwhasoo.

The cornerstone of the brand, Sulwhasoo Hair Treatment Oil is extracted from the seeds of the camellia tree, a traditional component of hair care in Korea and across Asia, to help hair grow, boost its shine, and fortify the scalp. Courtesy of Sulwhasoo.
Products from the brand Ginseng Sammi, launched in 1973. Courtesy of Sulwhasoo.

The beauty of Joseon white porcelain is reinterpreted from the viewpoint of digital civilization. When the present is interlinked with the past, the traditional genre of white porcelain is transformed into an unfamiliar but modern style that does away with three-dimensionality.
© Kang Suk-young, Ogh Sang-sun.

The camellia flower blooms in winter and symbolizes the integrity of women. Courtesy of Sulwhasoo.

The pine is the most beloved tree in Korea, symbolizing dignity and wisdom. Courtesy of Sulwhasoo.

Ginseng field. Ginseng is classified as one of the best medicinal ingredients in Korean herbal medicine. It is known to detoxify the body and refine the skin, and so brides in the Joseon dynasty prepared for their weddings by bathing in it. Courtesy of Sulwhasoo.
Products from the Snowise line's whitening range. Courtesy of Sulwhasoo.

Rocks at Seogwipo, Jeju-do, in South Korea. © Bae Bien-u.

Sulwhasoo employs many different botanical and herbal ingredients in its skin care products, including: top row, from left: Adhesive rehmannia, Satsuma mandarin, mountain peony, ladybells; middle row: Kobon, white lily, small Solomon's seal, dwarf lilyturf; bottom row: Adlay millet, East Indian lotus, Ya Jiu Hua, Chinese peony. Courtesy of Sulwhasoo.

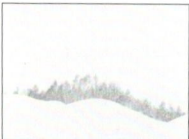

The meaning of Sulwhasoo is an arid and frail branch covered with snow-white flowers exploding in full blossom. The brand's name implies the blooming of snow flowers, as a metaphor for the renewal of aging skin. Korean women are very protective of the whiteness of their skin.
© Min Byung-hun.

Revitalizing Serum refines the skin and improves skin density; contains extracts of the Machilus tree. Courtesy of Sulwhasoo.
Sulwhasoo spa treatments respect tradition by using cool jade to heighten the effects of ginseng. © Réunion des Musées Nationaux/Art Resource, New York.

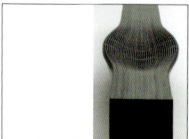

Artist Bae Se-hwa, whose work represents organic structure and formation, finds his aesthetic ideals in the harmony of nature, and believes that human beings can spiritually communicate with nature. He pursues the most tranquil structure, here presenting an artwork based on his steam project.
© Bae Se-hwa, Ogh Sang-sun.

Sulwhasoo Balancing Water and Balancing Emulsion. Courtesy of Sulwhasoo.
Sulwhasoo's flagship product, First Care Serum, formulated with Korean medicinal herbs, prepares skin to receive treatment and boosts the efficacy of skin care used throughout the regimen. Courtesy of Sulwhasoo.

© 2011 Assouline Publishing
601 West 26th Street, 18th floor
New York, NY 10001, USA
Tel.: 212 989-6810   Fax: 212 647-0005
www.assouline.com
Printed in China.
ISBN: 978-1-61428-012-5